Security Of1
TERRORISM RESPONSE GUIDE

Paul M. Maniscalco   **Hank T. Christen**

JONES AND BARTLETT PUBLISHERS

Sudbury, Massachusetts

BOSTON   TORONTO   LONDON   SINGAPORE

World Headquarters
Jones and Bartlett Publishers
40 Tall Pine Drive
Sudbury, MA 01776
978-443-5000
info@jbpub.com
www.jbpub.com

Jones and Bartlett Publishers
Canada
6339 Ormindale Way
Mississauga, Ontario L5V 1J2
CANADA

Jones and Bartlett Publishers
International
Barb House, Barb Mews
London W6 7PA
UK

Jones and Bartlett's books and products are available through most bookstores and online booksellers. To contact Jones and Bartlett Publishers directly, call 800-832-0034, fax 978-443-8000, or visit our website www.jbpub.com.

Substantial discounts on bulk quantities of Jones and Bartlett's publications are available to corporations, professional associations, and other qualified organizations. For details and specific discount information, contact the special sales department at Jones and Bartlett via the above contact information or send an email to specialsales@jbpub.com.

Production Credits
Chief Executive Officer: Clayton Jones
Chief Operating Officer: Don W. Jones, Jr.
President, Higher Education and Professional Publishing: Robert W. Holland, Jr.
V.P., Sales and Marketing: William J. Kane
V.P., Design and Production: Anne Spencer
V.P., Manufacturing and Inventory Control: Therese Connell
Publisher–Public Safety Group: Kimberly Brophy
Associate Managing Editor: Jennifer Reed
Production Editor/Composition/Text and Cover Design: Anne Spencer
Director of Marketing: Alisha Weisman
Senior Photo Researcher: Kimberly Potvin
Photo Researcher: Christine McKeen
Cover Image: © Chris Rossi/NNS/Landov
Chapter Opening Image: © Bobby Yip/Reuters/Landov
Printing and Binding: Malloy, Inc.
Cover Printing: Malloy, Inc.

Library of Congress Cataloging-in-Publication Data
Maniscalco, Paul M.
  Security officer's terrorism response guide / Paul M. Maniscalco and Hank T. Christen.
    p. cm.
  ISBN-13: 978-0-7637-3981-2
  ISBN-10: 0-7637-3981-2
  ISBN-13: 978-0-7637-4107-5
  ISBN-10: 0-7637-4107-8
  1. Terrorism investigation. 2. Incident command systems. 3. Emergency management. I. Christen, Hank T. II. Title.
  HV8079.T47M36 2006
  363.3202'4363289--dc22
                          2005029508
6048

Printed in the United States of America
10 09 08 07 06 10 9 8 7 6 5 4 3 2 1

Contents

The National Incident Management System

Incident Command System Organization

The Incident Command System (ICS) organizational structure is tactically initiated by the first arriving security officer in a modular fashion. It is based on the size and complexity of the incident, as well as the specifics of the hazardous environment created by the incident. When needed, separate functional elements can be established, each of which may be further

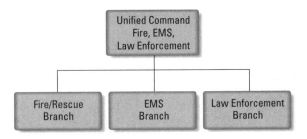

FIGURE 1-1 Simple ICS structure.

subdivided to enhance internal organizational management and external coordination (**FIGURE 1-1**).

NIMS

In 2003, the President directed the Secretary of Homeland Security to develop and administer a National Incident Management System (NIMS). This system provides a consistent nationwide template to enable federal, state, and local government and private-sector and nongovernmental organizations to work together effectively and efficiently to prepare for, prevent, respond to, and recover from domestic incidents, regardless of cause, size, or complexity, including acts of catastrophic terrorism.

The NIMS is a comprehensive national approach to the incident command system. The incident command system is structured so that there is a single authority or unified command team with overall responsibility to manage the incident. This person is identified as the incident commander. The incident commander or unified commander usually remains at a command post, the designated field command center. A field command center is typically a vehicle or building at the scene where the incident commander establishes an "office."

The NIMS will also improve coordination and cooperation between the public and private sector in a variety of domestic incidents, including:

- Acts of terrorism
- Wildland and urban fires
- Floods
- Hazardous materials spills
- Nuclear accidents

- Aircraft accidents
- Earthquakes
- Hurricanes
- Tornadoes
- Typhoons
- War-related disasters

To provide for interoperability and compatibility among federal, state, and local government, and the private sector, the NIMS includes a core set of concepts, principles, terminology, and technologies addressing the following:

- The incident command system
- Multiagency coordination systems
- Unified command
- Training
- Identification and management of resources
- Qualifications and certification
- Collection, tracking, and reporting of incident information and incident resources

While most incidents are regularly handled on a daily basis by a single local level department, there are important cases in which successful incident management operations depend on the involvement of multiple departments, levels of government, and the private sector. These cases require effective and efficient coordination across this broad spectrum of organizations and activities.

The NIMS includes several components that work together as a system to provide a national framework for preparing for, preventing, responding to, and recovering from domestic incidents. These components include the following:

1. **Command and management**—The NIMS standardizes incident management for all hazards and across all levels of government. The NIMS standard incident command structures are based on three key constructs: incident command system; multiagency coordination systems; and public information systems.

2. **Preparedness**—The NIMS establishes specific measures and capabilities that jurisdictions and agencies should develop and incorporate into an overall system to enhance operational prepared-

ness for incident management.

3. **Resource management**—The NIMS defines standardized mechanisms to describe, inventory, track, and dispatch resources before, during, and after an incident; it also defines standard procedures to recover equipment once it is no longer needed for an incident.

4. **Communications and information management**—Effective communications, information management, and information and intelligence sharing are critical aspects of domestic incident management. The NIMS communications and information systems ensure clear and rapid communication across all levels of government and departments.

5. **Supporting technologies**—The NIMS promotes national standards and interoperability for supporting technologies.

6. **Ongoing management and maintenance**—The Department of Homeland Security has established a NIMS Integration Center. This center provides strategic direction for and oversight of the NIMS, supporting routine maintenance and continuous improvement of the system in the long-term.

There are five major functions of the incident command system under the NIMS (**FIGURE 1-2**):

- **Command:** Responsible for the entire incident; this is the only function that is always staffed.
- **Operations:** Responsible for the actions taken at most emergency functions.
- **Planning:** Responsible for developing an Incident Action Plan to deal with the emergency incident.
- **Logistics:** Responsible for obtaining the resources needed to support the incident.
- **Finance/Administration:** Responsible for tracking expenditures and managing the administrative functions at the incident.

The Command Staff assists the Incident Commander or Unified Commander at the emergency incident:

- **Safety Officer:** Responsible for overall safety of the incident; has the authority to stop any action or operation if it creates a safety hazard.
- **Liaison Officer:** Responsible for coordinating operations between the agencies that are involved in the incident.

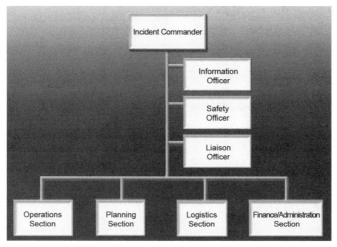

ICS Organizational Chart

Incident Commander

Information Officer

Safety Officer

Liaison Officer

Operations Section

Planning Section

Logistics Section

Finance/Administration Section

FIGURE 1-2 The ICS organizational chart.

- **Information Officer:** Responsible for coordinating media activities and providing the necessary information to the various media organizations.

NIMS Operations

The Operations Section is a key entity in the ICS. Operations functions are directed by the Operations Section Chief who is supervised by the Incident Commander or Unified Commander. Operations functions are further divided into branches as follows:

1. Law Enforcement
2. Fire/Rescue
3. Emergency Medical Services (EMS)
4. Public Works

Security officers will most likely be assigned to the Law Enforcement Branch of the Operations Section. The lead security officer or supervisor must coordinate with the Law Enforcement Branch Director. Security officer's duties vary according to the incident. In general, these duties may include:

1. Entry and exit control

2. Perimeter control

3. Evacuation operations

4. Vehicle, property, and equipment security

Summary

The NIMS can be used by all levels of government and private sector agencies to deal with incidents both large and small in scale. All emergency incident responders, from the head of FEMA to the facilities security officer, will be trained in the NIMS protocol. The NIMS will ensure that all levels of government and all types of agencies will work together efficiently and safely.

2

The Homeland Security Advisory System

- The Homeland security alert level should be briefed at each shift change.
- Notify critical personnel when the alert level is escalated.
- Initiate appropriate procedures in the organization emergency plan when alert levels change.

Introduction

The U.S. Department of Homeland Security (DHS), in conjunction with the Homeland Security Council, has developed and implemented the Homeland Security Advisory System (**FIGURE 2-1**). This system is designed to provide quick, comprehensive information concerning the potential threat of terrorist attacks or threat levels. Threat conditions can apply nationally, regionally, by industrial sector, or by specific target.

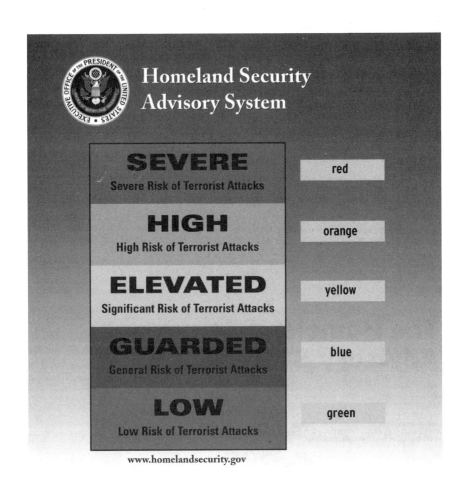

FIGURE 2-1 Homeland Security Advisory System.

Pre-Incident Activities: Low Condition (Green)

This condition is declared when there is a low risk of terrorist attacks. Agencies should consider the following general measures in addition to any agency-specific protective measures they develop and implement:

1. Refine and exercise pre-planned protective measures.
2. Ensure that personnel receive proper training on the Homeland Security Advisory System and on specific pre-planned departmental or agency protective measures.
3. Institute a process to ensure that all facilities and regulated sectors are regularly assessed for vulnerability to terrorist attacks, and that all reasonable measures are taken to mitigate these vulnerabilities.

Advise organization to:

- Develop an organization emergency plan, including meeting locations in the event of an attack, and a contact point, such as separate local or out-of-state facilities.
- Keep recommended immunizations up to date.
- Know how to turn off power, gas, and water service to facilities. Keep the proper tools available to do so.
- Know what hazardous materials are stored in facilities.
- Support the efforts of local emergency responders (fire fighters, law enforcement officers, and emergency medical service personnel).
- Know the natural hazards that are prevalent in your area, and what measures can be taken to mitigate these hazards.
- Maintain a medical plan.

Pre-Incident Activities: Guarded Condition (Blue)

This condition is declared when there is a general risk of terrorist attacks. *In addition to taking the protective measures listed under the previous threat condition*, agencies should consider the following general measures as well as the agency-specific protective measures:

1. Check communications with designated emergency response or command locations.
2. Review and update emergency response procedures.
3. Provide organization with any information that would strengthen its ability to act appropriately.

Advise organization to:

- Continue normal activities, but be watchful and report suspicious activities to security officers.
- Review organization emergency plans, including meeting locations and distant points of contact in the event of an attack.
- Be familiar with the local natural and technological (man-made) hazards in your community and know what measures can be taken to mitigate these hazards.
- Maintain a medical plan.
- Increase organizational emergency preparedness, provide for back-up power, and check critical supply caches.
- Monitor local and national news for terrorist alerts.

Pre-Incident Activities: Elevated Condition (Yellow)

This condition is declared when there is a significant risk of terrorist attacks. *In addition to the protective measures listed under the previous threat conditions*, agencies should consider the following general measures as well as the protective measures that they will develop and implement:

1. Increase surveillance of critical locations.
2. Coordinate emergency plans as appropriate with nearby jurisdictions.
3. Address whether the precise characteristics of the threat require the further refinement of pre-planned protective measures.
4. Implement, as appropriate, contingency and emergency response plans.

Advise organization to:

- Continue normal activities, but report suspicious activities to security officers or local law enforcement agencies.
- Maintain a medical plan.
- Network with similar organizations and local public safety agencies for mutual support during a disaster or terrorist attack. Update previously established organization emergency plans, including a distant point of contact. Update preparations to shelter in place, such as preparing of materials to seal openings in facilities and secure air conditioning and air handling systems.
- Know the critical facilities and specific hazards located in your facility and report suspicious activities at or near these sites.
- Monitor media reports concerning terrorism situations.

Pre-Incident Activities: High Condition (Orange)

This condition is declared when there is a high risk of terrorist attacks. *In addition to the protective measures listed under the previous threat conditions*, agencies should consider the following general measures as well as agency-specific protective measures:

1. Coordinate necessary security efforts with federal, state, and local law enforcement agencies or any National Guard or other appropriate security organizations.

2. Take additional precautions at public events and consider alternative venues or even cancellation.

3. Prepare to execute contingency procedures, such as moving to an alternate site or dispersing a workforce.

4. Restrict threatened facility access to essential personnel only.

Advise organization to:

- Expect some delays, baggage searches, and restrictions as a result of heightened security at buildings and facilities.

- Continue to monitor world and local events as well as local government threat advisories.

- Report suspicious activities at or near critical facilities to local law enforcement agencies.

- Avoid leaving unattended packages or briefcases in public areas.

- Inventory and organize emergency supply caches and discuss emergency plans with organizational departments. Re-evaluate emergency procedures based on any specific threat. Identify distant points of contact. Ensure that supplies for sheltering in place are maintained. Initiate emergency medical and fire response plans.

- Advise all personnel to take reasonable personal security precautions. Be alert to your surroundings, avoid placing yourself in a vulnerable situation.

Pre-Incident Activities: Severe Condition (Red)

This condition reflects a severe risk of terrorist attacks. Under most circumstances, the protective measures for a Severe Condition are not intended to be sustained for extended periods of time. *In addition to the protective measures listed under the previous threat conditions*, agencies should consider the following general measures as well as the agency-specific protective measures:

1. Increase or redirect personnel to address critical emergency needs.

2. Assign emergency response personnel and pre-position and mobilize specially-trained teams or resources.

3. Recall all available security personnel.

4. Close all facilities—initiate lockdown procedures.

5. Require an identification check of all on-site personnel and search all vehicles attempting entry to critical facilities.

Advise organization to:

- Report suspicious activities and call for immediate assistance.
- Expect delays, searches of purses and bags, and restricted access to facilities.
- Expect traffic delays and restrictions.
- Take personal security precautions (such as avoiding target facilities) to avoid becoming a victim of a crime or a terrorist attack.
- Do not travel into areas affected by an attack.
- Keep emergency supplies accessible and your automobile's fuel tank full.
- Initiate highest level of emergency medical readiness.
- Be prepared to evacuate facilities or to shelter in place on the order of local authorities. Initiate emergency supply procedures.
- Be suspicious of persons taking photographs of critical facilities, asking detailed questions about physical security, or who are dressed inappropriately for weather conditions (suicide bombers). Report these incidents immediately to law enforcement.
- Closely monitor news reports and Emergency Alert System (EAS) radio/TV stations.
- Avoid passing unsubstantiated information.

Source: Homeland Security Presidential Directive #3, Washington, DC; March 12, 2002; and *Homeland Security Advisory System Recommendations for Individuals, Families, Neighborhoods, Schools, and Businesses*, The American Red Cross, www.redcross.org, March 2003.

Potential Terrorist Targets and Tactics

TACTICAL ACTIONS

- Identify critical infrastructure within your organization as part of the organization's emergency plan.
- Monitor the media and internal/external intelligence reports to maintain a situational awareness of major incidents in the local, regional, state, and national arena.
- Consider key leaders or executives as part of the critical infrastructure.
- Protect and secure major events, ceremonies, and festivals.

Introduction

Terrorists are usually motivated by a cause and choose targets they believe will help them achieve their goals and objectives. Many terrorist incidents aim to instill fear and panic among the general population and to disrupt daily ways of life. Sometimes the objective is to sabotage or to destroy a facility that is significant to the terrorist cause. The ultimate goal could be to cause economic turmoil by interfering with transportation, trade, or commerce.

Terrorists choose a method of attack they think will make the desired statement or achieve the maximum results. They may vary their methods or change them over time. Explosive devices have been used in thousands of terrorist attacks; in recent years, however, there has been a significant increase in the number of suicide bombings.

Many terrorist incidents in the 1980s involved the taking of hostages on hijacked aircraft or cruise ships. Sometimes, only a few people were held; other times, hundreds of people were taken hostage. Diplomats, journalists, and athletes were targeted in several incidents, and the terrorists often offered to release their hostages in exchange for the release of imprisoned individuals allied with the terrorist cause. More recently terrorist actions have endangered thousands of lives with no warning or request to negotiate.

Terrorism can occur in any community. A rural ski lodge, tucked away in the mountains, may be attacked by an environmental group that is upset with plans for expansion. A small retailer in an upscale suburban community could become the target of an animal rights group that objects to the sale of fur coats. An anti-abortion group might plant a bomb at a local community health clinic. There are many causes with supporters who range from peaceful, nonviolent organizations to fanatical fringe groups. Being highly familiar with your work environment will greatly assist the security officer with truly understanding prevailing threat and targeting potential.

Ecoterrorism

Ecoterrorism refers to illegal acts committed by groups supporting environmental or related causes. Examples include spiking trees to sabotage logging operations, vandalizing a university research laboratory that is conducting experiments on animals, or firebombing a store that sells fur coats. The majority of domestic ecoterrorism incidents have been attributed to special interest groups such as the Earth Liberation Front (ELF) and the Animal Liberation Front (ALF) or groups claiming affiliation to them or their agenda.

Infrastructure Targets

Most of the critical infrastructure in the United States is a public utility or private sector entity. Terrorists might strike bridges, tunnels, or subways in an attempt to disrupt transportation and inflict a large number of casualties. They could also attack the public water supply or try to disable the electrical power distribution system, telephones, or the Internet. Disruption of a community's 9-1-1 system or public safety radio network would have a very direct and negative impact on emergency response agencies.

Symbolic Targets

Monuments such as the Lincoln Memorial, Washington Monument, or Mount Rushmore may be targeted by groups who want to attack symbols of national pride and accomplishment. Foreign embassies and institutions might be attacked by groups promoting revolution within those countries or protesting their international policies. Religious institutions or other visible icons are potential targets of hate groups. Attacks can also be planned to affect the economy by targeting manufacturing, banking, financial institutions, or retail shopping centers. By targeting these symbols, terrorist groups seek to make people aware of their demands and create a sense of fear in the public.

Civilian Targets

Terrorists who attack civilian targets such as shopping malls, schools, or stadiums indiscriminately kill or injure the maximum number of potential victims. This list also includes high-profile businesses and organizations. Their goal is to create fear in every member of society and to make citizens feel vulnerable in their daily lives. Letter bombs or letters that contain a biologic agent have a similar effect.

Cyberterrorism

Groups might engage in cyberterrorism by electronically attacking government or private computer systems. Several attempts have been made to disrupt the Internet or to attack government computer systems and other critical networks.

Agroterrorism

Agroterrorism could include the use of chemical or biologic agents to attack the agricultural industry or the food supply. The introduction of a disease such as foot-and-mouth to the livestock population could result in

Table 3-1	Potential Terrorist Targets

Ecoterrorism Targets
- Controversial development projects
- Environmentally sensitive areas
- Research facilities
- See also infrastructure targets
- Shopping centers

Infrastructure Targets
- Bridges and tunnels
- Emergency facilities
- Hospitals
- Oil refineries
- Pipelines and fuel storage
- Power plants and electrical distribution systems
- Railroads, ports, and airports
- Telecommunications systems
- Water reservoirs and treatment plants

Symbolic Targets
- Embassies
- Government buildings
- Military bases
- National monuments and historic sites
- Places of worship
- Banking/Financial institution
- Shopping/Retail

Civilian Targets
- Arenas and stadiums
- Airports and railroad stations
- Mass-transit systems
- Schools and universities
- Shopping centers
- Theme parks

Cyberterrorism Targets
- Banking and finance computer systems
- Business computer systems
- Court computer systems
- Government computer systems
- Public safety dispatch and computer systems
- Military computer systems

Agroterrorism Targets
- Crops
- Feed storage
- Grain elevators
- Livestock and poultry

4

Chemical Terrorism

TACTICAL ACTIONS

- Call for help—describe device and/or symptoms of victims.
- Secure the area using recommended distances.
- Establish a safe zone that is uphill and upwind.
- Avoid exposure—wear proper protective equipment.
- Control scene entry until law enforcement arrives.
- Assist in scene security or evacuation if requested by law enforcement.

Introduction

Chemical agents are manmade substances that can have devastating effects on living organisms. They can be produced in liquid, powder, or vapor form depending on the desired route of exposure and distribution technique. These agents have been implicated in thousands of deaths since being introduced on the battlefield in World War I. Since then have been used to terrorize civilian populations. These agents include:

- Vesicants (blister agents)
- Respiratory agents (choking agents)
- Nerve agents
- Metabolic agents (blood agents)
- Common industrial chemicals

Background

Chemical agents are liquids or gases that are dispersed to kill or injure. Modern-day chemicals were first developed during World War I and World War II. During the Cold War, many of these agents were perfected and stockpiled. While the United States has long renounced the use of chemical weapons, many nations still develop and stockpile them. These agents are deadly and pose a threat if acquired by terrorists. There are thousands of industrial chemicals shipped and stored daily throughout the United States.

Chemical weapons have several classifications. The properties or characteristics of an agent can be described as liquid, gas, or solid material. Route of exposure is a term used to describe how the agent most effectively enters the body. Chemical agents can have either a vapor or contact hazard. Agents with a vapor hazard enter the body through the mouth, nose, and lungs in the form of vapors. Agents with a contact hazard (or skin hazard) give off very little vapor or no vapors and enter the body through the skin.

Vesicants (Blister Agents)

The primary route of exposure of blister agents, or vesicants, is the skin (contact); however, if vesicants are left on the skin or clothing long enough, they produce vapors that can enter the lungs. Vesicants cause burn-like blisters to form on the victim's skin as well as in the lungs. The vesicant agents consist of sulfur mustard (H), Lewisite (L), and phosgene oxime (CX) (the symbols H, L, and CX are military designations for these chemicals). The vesicants usually cause the most damage to damp or moist areas of the body, such as the armpits, groin, and lungs. Signs of vesicant exposure on the skin include:

- Skin irritation, burning, and reddening
- Immediate intense skin pain (with L and CX)
- Formation of large blisters
- Gray discoloration of skin (a sign of permanent damage seen with L and CX)
- Swollen and closed or irritated eyes
- Permanent eye injury (including blindness)

If vapors were inhaled, the patient may experience the following:

- Hoarseness and stridor (harsh, high-pitched inspiratory sound)
- Severe cough
- Hemoptysis (coughing up of blood)
- Severe shortness of breath

Sulfur mustard (agent H) is a brownish, yellowish oily substance that is generally considered very persistent. When released, mustard has the distinct smell of garlic or mustard and is quickly absorbed into the skin and/or mucous membranes (nasal and throat passages). As the agent is absorbed into the skin, it begins an irreversible process of damage to the cells. Absorption through the skin or mucous membranes usually occurs within seconds, and damage to the underlying cells takes place within 1 to 2 minutes.

Mustard is considered a mutagen, which means that it mutates, damages, and changes the structures of cells. Eventually, cellular death will occur. On the surface, the patient will generally not produce any signs or symptoms until 4 to 6 hours after exposure (depending on concentration and amount of exposure).

The victim will develop a progressive reddening of the affected area, which will gradually develop into large blisters. These blisters are very similar in shape and appearance to those associated with thermal second-degree burns. The fluid within the blisters does not contain any of the agent; however, the skin covering the area is considered to be contaminated until decontamination by trained personnel has been performed.

Mustard also reduces the body's resistance to infections. It also damages the throat and lungs.

Lewisite (L) and phosgene oxime (CX) produce blister wounds very similar to mustard. They are highly volatile and have a rapid onset of symptoms. These agents produce immediate intense pain and discomfort when contact is made. The victim may have a grayish discoloration at the contaminated site. These agents damage the victim's body tissue, but do not weaken the body's overall resistance to infection.

Vesicant Agent Treatment

There are no antidotes for mustard or CX exposure. BAL (British Anti-Lewisite) is the antidote for agent L; however, it is not carried by civilian EMS. The EMS responder must ensure that the victim has been decontaminated before care is initiated. The victim may require help breathing if any agent has been inhaled, but this should not occur until after decontamination. Call EMS immediately to care for these victims.

Pulmonary Agents (Choking Agents)

The pulmonary agents are gases that cause immediate harm to persons exposed to them. The primary route of exposure for these agents is through the nose, throat, and lungs which makes them an inhalation or vapor hazard. Once inside the lungs, they damage the lung tissue and fluid leaks into the lungs. The victim develops difficulty breathing. These agents produce respiratory-related symptoms such as shortness of breath, rapid respirations, and fluid in the lungs. This class of chemical agents consists of chlorine (CL) and phosgene.

Chlorine (CL) was the first chemical agent ever used in warfare. It has a distinct odor of bleach and creates a green haze when released as a gas. Initially it produces a choking sensation. The victim may later experience:

- Shortness of breath
- Chest tightness
- Hoarseness and stridor
- Gasping and coughing

With serious exposures, patients may experience fluid in the lungs, complete airway constriction, and death. The fumes from a mixture of household bleach (CL) and ammonia create an acid gas that produces similar effects. Each year, such mixtures overcome hundreds of people when they try to mix household cleaners.

Phosgene should not be confused with phosgene oxime, a blistering agent, or vesicant. Not only has phosgene been produced for chemical warfare, but it is a product of combustion such as might be produced in a fire at a textile factory or house, or from metalwork or burning Freon (a liquid chemical used in refrigeration). Therefore, you may encounter a victim of exposure to this gas. Phosgene is a very potent agent that has a delayed onset of symptoms, usually hours. Unlike CL, when phosgene enters the body, it generally does not produce severe irritation, which would possibly cause the victim to leave the area or hold his or her breath. In fact, the odor produced by the chemical is similar to that of freshly mown grass or hay. The result is that much more of the gas is allowed to enter the body unnoticed. The initial symptoms of a mild exposure may include:

- Nausea
- Chest tightness
- Severe cough
- Shortness of breath upon exertion

The victim of a severe exposure may present with shortness of breath at rest, and excessive fluid in the lungs (the victim will cough up large amounts of fluid).

Pulmonary Agent Treatment

The best initial treatment for any pulmonary agent is to remove the victim from the contaminated atmosphere. This should be done by trained personnel in the proper personal protective equipment (PPE). Call EMS immediately.

Nerve Agents

Nerve agents are among the most deadly chemicals developed. Designed to kill large numbers of people with small quantities, nerve agents can cause cardiac arrest within seconds to minutes of exposure.

G agents came from the early nerve agents, the G series, which were developed by German scientists (hence the G) in the period after World War I and into World War II. There are three G series agents:

- **Sarin (GB):** Highly volatile colorless and odorless liquid. Turns from liquid to gas within seconds to minutes at room temperature. Highly lethal. Sarin is primarily a vapor hazard, with the mouth, nose, and lungs as the main routes of entry. This agent is especially dangerous in enclosed environments such as office buildings, shopping malls, or subway cars. When this agent comes into contact with skin, it is quickly absorbed and evaporates. When sarin is on clothing, it has the effect of off-gassing, which means that the vapors are continuously released over a period of time (like perfume). This renders the victim as well as the victim's clothing contaminated.

- **Soman (GD):** Twice as persistent as sarin and five times as lethal. It has a fruity odor as a result of the type of alcohol used in the agent and generally has no color. This agent is both a contact and inhalation hazard that can enter the body through skin absorption and through breathing. A unique additive in GD causes it to bind to the cells that it attacks faster than any other agent. This irreversible binding is called aging, which makes it more difficult to treat victims who have been exposed.

- **TABUN (GA):** Approximately half as lethal as sarin; under the proper conditions it will remain for several days. It also has a fruity smell and an appearance similar to sarin. The components used to manufacture GA are easy to acquire and the agent is easy to manufacture, which make it unique. GA is both a contact and inhalation hazard that can enter the body through skin absorption as well as through breathing.

- **V agent (VX):** Clear, oily agent that has no odor and looks like baby oil. V agent was developed by the British after World War II and has similar chemical properties to the G series agents. The difference is that VX is over 100 times more lethal than sarin. VX is so persistent that given the proper conditions it will remain relatively unchanged for weeks to months. These properties make VX primarily a contact hazard, because it lets off very little vapor. It is easily absorbed into the skin, and the oily residue that remains on the skin's surface is extremely difficult to decontaminate.

Nerve agents all produce similar symptoms, but have varying routes of entry. Nerve agents differ slightly in lethal concentration or dose and also differ in their volatility. Some agents are designed to become a gas quickly, while others remain liquid for a period of time. These agents have been used successfully in warfare and to date represent the only type of chemical agent that has been used successfully in a terrorist act. Once the agent has entered the body through skin contact or through the respiratory system, the victim will begin to exhibit a pattern of predictable symptoms. Like all chemical agents, the severity of the symptoms will depend on the route of exposure and the amount of agent to which the victim was exposed. The resulting symptoms are described below using the military mnemonic SLUDGEM and the medical mnemonic DUMBELS (**TABLE 4-1**). SLUDGEM/DUMBELS mnemonics are used to describe the symptoms of nerve agent exposure. The medical mnemonic is more useful to you because it lists the more dangerous symptoms associated with exposure to nerve agents.

Table 4-1 Symptoms of Persons Exposed to Nerve Agents

Military Mnemonic: SLUDGEM

Salivation, Sweating

Lacrimation (excessive tearing)

Urination

Defecation, Drooling, Diarrhea

Gastric upset and cramps

Emesis (vomiting)

Muscle twitching

Medical Mnemonic: DUMBELS

Diarrhea

Urination

Miosis (pinpoint pupils)

Bradycardia (slow heart beat), Bronshospasm (spasm of the bronchioles in the lungs)

Emesis (vomiting)

Lacrimation (excessive tearing)

Seizures, Salivation, Sweating

Nerve Agent Treatment

Call EMS. Fatalities from severe exposure occur as a result of breathing complications, which lead the victim to stop breathing. Once the victim has been decontaminated, EMS will treat these victims aggressively. There is an antidote for nerve agent exposure.

TABLE 4-2 has been provided for quick reference and comparison of the nerve agents.

Table 4-2	The Nerve Agents					
Name	Military Designation	Odor	Special Features	Onset of Symptoms	Volatility	Route of Exposure
Tabun	GA	Fruity	Easy to manufacture	Immediate	Low	Both contact and vapor hazard
Sarin	GB	None (if pure or strong)	Will off-gas while on victim's clothing	Immediate	High	Primarily respiratory vapor hazard; extremely lethal if skin contact is made
Soman	GD	Fruity	Ages rapidly, making it difficult to treat	Immediate	Moderate	Contact with skin; minimal vapor hazard
V agent	VX	None	Most lethal chemical agent; difficult to decontaminate	Immediate	Very low	Contact with skin; no vapor hazard (unless aerosolized)

Metabolic Agents (Cyanides)

Hydrogen cyanide (AC) and cyanogen chloride (CK) are both agents that affect the body's ability to use oxygen. Cyanide is a colorless gas that has an odor similar to almonds. The effects of the cyanides begin on the cellular level and are very rapidly seen throughout the body. Beside the nerve agents, metabolic agents are the only chemical weapons known to kill within seconds to minutes. Unlike nerve agents, however, these deadly gases are commonly found in many industrial settings. Cyanides are produced in massive quantities throughout the United States every year for industrial uses such as gold and silver mining, photography, lethal injections, and plastics processing. They are often present in fires associated with textile or plastic factories. In fact, cyanide is naturally found in the pits of many fruits in very low doses. There is very little difference in the symptoms found between AC and CK. In low doses, these chemicals are associated with dizziness, light-headedness, headache, and vomiting. Higher doses will produce symptoms that include:

- Shortness of breath and gasping respirations
- Rapid breathing
- Flushed skin color
- Rapid heart beat
- Altered mental status
- Seizures
- Coma
- Absence of breathing
- Cardiac arrest

The symptoms associated with the inhalation of a large amount of cyanide will all appear within several minutes. Death is likely unless the victim is treated promptly. Call EMS.

Cyanide Agent Treatment

Call EMS. Several medications act as antidotes, but most EMS services do not carry them. Once trained personnel in the proper PPE have removed the victim from the source of exposure, even if there is no liquid contamination, all of the victim's clothes must be removed to prevent off-gassing. Trained and protected personnel must decontaminate any victims who may have been exposed to liquid contamination before EMS can initiate treatment.

Industrial Chemicals/Insecticides

The basic chemical ingredient in nerve agents is organophosphate. This is a common chemical that is used in lesser concentrations for insecticides. While industrial chemicals do not possess sufficient lethality to be effective weapons of mass destruction, they are easy to acquire, inexpensive, and would have similar effects as the nerve agents. Crop-duster planes could be used to disseminate these chemicals. Be cautious where insecticide equipment is stored and used, such as a farm or supply store that sells these products. Beware of water runoff from an organophosphate incident. The symptoms and medical management of victims of organophosphate insecticide poisoning are identical to those of the nerve agents.

Common industrial chemicals can be intentionally released—they become chemical weapons. These chemicals can be in liquid or gaseous form. Examples of common chemical hazards include:

1. Flammability—gasoline, diesel, or propane fuel
2. Corrosive/caustic—sulphuric acid or sodium hydroxide
3. Respiratory—chlorine gas or ammonia

Security officers may be able to identify the type of chemical by obtaining shipping papers or by reading a placard through binoculars from a safe distance. Maintain a safe distance from the vehicle or container. Stay uphill and upwind from the chemical source.

Summary

Table 4-3 summarizes the chemical agents. The odors of the particular chemicals are provided for informational purposes only. The sense of smell is a poor tool to use to determine whether there is a chemical agent present. Many persons are unable to smell the agents, and the odor could be derived from another source. This information is useful to you if you receive reports from victims claiming to smell of bleach or garlic, for example. You should never enter a potentially hazardous area and "smell" to determine whether a chemical agent is present.

Table 4-3 Chemical Agents

Name	Military Designation	Odor	Lethality	Onset of Symptoms	Volatility	Primary Route of Exposure
Nerve agents	Tabun (GA); Sarin (GB); Soman (GD); VX	Fruity or none	Most lethal chemical agents; can kill within minutes; effects are reversible with antidotes	Immediate	Moderate (GA, GD); Very high (GB); Low (VX)	GA-both; GB-vapor hazard; GD-both; VX-contact hazard
Vesicants	Mustard (H); Lewisite (L); Phosgene oxime (CX)	Garlic (H); Geranium (L)	Causes large blisters to form on victims; may severely damage upper airway if vapors are inhaled; severe, intense pain and grayish skin discoloration (L and CX)	Delayed (H); Immediate (L, CX)	Very low (H, L); Moderate (CX)	Primarily contact with some vapor hazard
Pulmonary agents	Chlorine (CL); Phosgene (CG)	Bleach (CL); Cut grass (CG)	Causes irritation choking (CL); severe pulmonary edema (CG)	Immediate (CL); Delayed (CG)	Very high	Vapor hazard
Cyanide agents	Hydrogen cyanide (AC); Cyanogen chloride (CK)	Almonds (AC); Irritating (CK)	Highly lethal chemical gases; can kill within minutes; effects are reversible with antidotes	Immediate	Very high	Vapor hazard

Biological Terrorism

TACTICAL ACTIONS

- The effects of a biological attack are not immediate—symptoms may take days or weeks to appear.
- Report suspicious powder incidents to local response agencies immediately.
- Avoid direct contact with infected individuals—stay isolated from body fluids such as blood, urine, vomitus, sputum, and fecal matter.
- Security officers must use appropriate protective equipment if contact with an infected individual is necessary.
- Secure an area where biological contamination is suspected.
- Turn off air conditioning and air handling systems in contaminated facilities.
- Exposed security officers should undergo immediate medical evaluation.

Introduction

Biological agents are organisms that cause disease. They are generally found in nature; however, for terrorist use, they are cultivated, synthesized, and mutated in a laboratory. The weaponization of biological agents is performed to artificially maximize the target population's exposure to the germ, thereby exposing the greatest number of people and achieving the desired result.

The primary types of biological agents that you may come into contact with during a biological event include:

- Viruses
- Bacteria
- Toxins

Background

Biological agents pose many difficult issues when used as a weapon of mass destruction. Biological agents can be almost completely undetectable. Also, most of the diseases caused by these agents will be similar to other minor illnesses commonly seen.

Biological agents are grouped as viruses, bacterium, or neurotoxins and may be spread in various ways. Dissemination is the means by which a terrorist will spread the agent—for example, poisoning the water supply or aerosolizing the agent into the air or ventilation system of a building. A disease vector is an animal that spreads disease, once infected, to another animal. For example, the plague can be spread by infected rats, smallpox by infected persons, and West Nile virus by infected mosquitoes. How easily the disease is able to spread from one human to another human is called communicability. Some diseases, such as HIV, are difficult to spread by routine contact. Therefore communicability is considered low. In other instances when communicability is high, such as with smallpox, the person is considered contagious.

Incubation describes the period of time between the person becoming exposed to the agent and when symptoms begin. Although victims may not exhibit signs or symptoms, he or she may be contagious.

If the agent is in the form of a powder, such as in the October 2001 attacks involving anthrax powder mailed in letters, the incident must be handled by HazMat specialists. Victims who have come into direct contact with the agent need to be decontaminated before any EMS contact or treatment is initiated. Doctors are very concerned about the use of contagious diseases as weapons, because the resulting epidemic could overwhelm the healthcare system. There is growing concern that a terrorist group will use one of these agents against a civilian population.

Viruses

Viruses are germs that require a living host to multiply and survive. A virus is a simple organism and cannot thrive outside of a host (living body). Once in the body, the virus will invade healthy cells and replicate itself to spread through the host. As the virus spreads, so does the disease that it carries. Viruses survive by moving from one host to another.

Although some viral agents do have vaccines, there is no treatment for a viral infection other than antivirals for some agents. Because of this characteristic, the following viruses have been used as terrorist agents.

Smallpox

Smallpox is a highly contagious disease. If you suspect a victim has smallpox, call EMS. All forms of precautions must be used to prevent contamination. The last natural case of smallpox in the world was seen in 1977. Before the rash and blisters show, the illness will start with a high fever and body aches and headaches.

An easy, quick way to differentiate the smallpox rash from other skin disorders is to observe the size, shape, and location of the lesions. In smallpox, all the lesions are identical in their development. In other skin disorders, the lesions will be in various stages of healing and development. Smallpox blisters also begin on the face and extremities and eventually move toward the chest and abdomen. The disease is in its most contagious phase when the blisters begin to form. Unprotected contact with these blisters will cause contamination. There is a vaccine to prevent smallpox; however, it has been linked to medical complications, and in very rare cases, death. Vaccination against the disease is part of a national strategy to respond to a terrorist threat. Because the vaccine does have some risk, only first responders have been offered the vaccine. Should an outbreak occur, vaccine would be offered to all people at risk. **TABLE 5-1** shows characteristics of smallpox.

Table 5-1	Characteristics of Smallpox
Dissemination	Aerosolized for warfare or terrorist uses.
Communicability	High from infected individuals or items (such as blankets used by infected victims). Person-to-person transmission is possible.
Route of entry	Through inhalation of coughed droplets or direct skin contact with blisters.
Signs and symptoms	Severe fever, body aches, headaches, small blisters on the skin, bleeding of the skin and nasal passages. Incubation period is 10 to 12 days and the duration of the illness is approximately 4 weeks.

Viral Hemorrhagic Fevers

Viral hemorrhagic fevers (VHF) consist of a group of diseases that include the Ebola, Rift Valley, and Yellow Fever viruses, among others. This group of viruses causes the blood in the body to seep out from the tissues and blood vessels. Initially, the victim will have flu-like symptoms, progressing to more serious symptoms such as internal and external bleeding. Outbreaks are not uncommon in Africa and South America. Outbreaks in the United States, however, are extremely rare. Mortality rates can range from 5% to 90%, depending on the strain of virus, the victim's age and health condition, and the availability of a modern health care system (**TABLE 5-2**).

Bacteria

Unlike viruses, bacteria do not require a host to multiply and live. Bacteria are much more complex and larger than viruses and can grow up to 100 times larger than the largest virus. Bacteria contain all the cellular structures of a normal cell and are completely self-sufficient. Most importantly, bacterial infections can be fought with antibiotics.

Most bacterial infections will generally begin with flu-like symptoms, which make it quite difficult to identify whether the cause is a biological attack or a natural epidemic. Biological agents have been developed and used for centuries during times of war.

Table 5-2	Characteristics of Viral Hemorrhagic Fevers
Dissemination	Direct contact with an infected person's body fluids. It can also be aerosolized for use in an attack.
Communicability	Moderate from person to person, or contaminated items.
Route of entry	Direct contact with an infected person's body fluids.
Signs and symptoms	Sudden onset of fever, weakness, muscle pain, headache, and sore throat. All of these symptoms are followed by vomiting and as the virus runs it course, internal and external bleeding.

Inhalation and Cutaneous Anthrax (*Bacillus anthracis*)

Anthrax is a deadly bacteria that lays dormant in a spore (protective shell). When exposed to the optimal temperature and moisture, the germ will be released from the spore. The routes of entry for anthrax are inhalation, through the skin, or from consuming food that contain spores. The inhalational form or pulmonary anthrax is the most deadly and often presents as a severe cold. Pulmonary anthrax infections are associated with a 90% death rate if untreated. Antibiotics can be used to treat anthrax successfully. There is also a vaccine to prevent anthrax infections (**TABLE 5-3**).

Table 5-3	Characteristics of Anthrax
Dissemination	Aerosol
Communicability	Only when passed through the skin (rare).
Route of entry	Through inhalation of spore or skin contact with spore or direct contact with skin wound.
Signs and symptoms	Flu-like symptoms, fever, breathing distress, and breathing failure after 3 to 5 days of flu-like symptoms.

Plague (Bubonic/Pneumonic)

Of all the infectious diseases known to humans, none has killed as many as the plague. The 14th century plague that ravaged Asia, the Middle East, and finally Europe (the Black Death) killed an estimated 33 to 42 million people. Later on, in the early 19th century, almost 20 million in India and China perished due to plague. Infected rodents and fleas spread the plague. When a person is either bit by an infected flea or comes into contact with an infected rodent (or the waste of the rodent), the person can contract bubonic plague.

Bubonic plague infects the lymphatic system (a passive circulatory system in the body that bathes the tissues in lymph and works with the immune system). When this occurs, the patient's lymph nodes (area of the lymphatic system where infection-fighting cells are housed) become infected and grow. The glands of the nodes will grow large (up to the size of a tennis ball) and round, forming buboes. If left untreated, the infection may spread through the body, possibly leading to death. This form of plague is not contagious and is not likely to be seen in a bioterrorist incident.

Pneumonic plague is a lung infection, also known as plague pneumonia, that results from inhalation of plague bacteria. This form of the disease is contagious and has a much higher death rate than the bubonic form. This form of plague therefore would be easier to disseminate (aerosolized), has a higher death rate, and is contagious (**TABLE 5-4**).

Table 5-4	Characteristics of Plague
Dissemination	Aerosol.
Communicability	Bubonic: low, only from contact with fluid in buboe. Pneumonic: high, from person to person.
Route of entry	Ingestion, inhalation, or skin contact.
Signs and symptoms	Fever, headache, muscle pain and tenderness, pneumonia, shortness of breath, extreme lymph node pain and enlargement (bubonic).

Neurotoxins

Neurotoxins are the most deadly substances known to humans. The strongest neurotoxin is 15,000 times more lethal than VX and 100,000 times more lethal than sarin. These toxins are produced from plants, marine animals, molds, and bacteria. The route of entry for these toxins is through ingestion, inhalation from aerosols, or injection. Unlike viruses and bacteria, neurotoxins are not contagious and have a faster onset of symptoms. Although these biological toxins have immense destructive potential, they have not been used successfully as a weapon of mass destruction.

Botulinum Toxin

The most potent neurotoxin is botulinum, which is produced by bacteria. When introduced into the body, this neurotoxin affects the nervous system's ability to function. Voluntary muscle control will diminish as the toxin spreads. Eventually the toxin will cause muscle paralysis that begins at the head and face and travels downward throughout the body. The victim will stop breathing (**TABLE 5-5**).

Ricin

While not as deadly as botulinum, ricin is still five times more lethal than VX. This toxin is derived from mash that is left from the castor bean. When introduced into the body, ricin causes fluid in the lungs and breathing and heart failure leading to death.

Table 5-5: Characteristics of Botulinum Toxin

Dissemination	Aerosol or food supply sabotage or injection.
Communicability	None.
Route of entry	Ingestion or gastrointestinal.
Signs and symptoms	Dry mouth, stomach blockage, urinary retention, constipation, nausea and vomiting, abnormal pupil dilation, blurred vision, double vision, drooping eyelids, difficulty swallowing, difficulty speaking, and breathing failure due to paralysis.

The clinical picture depends on the route of exposure. The toxin is quite stable and extremely toxic by many routes of exposure, including inhalation. Perhaps 1 to 3 mg of ricin can kill an adult, and the ingestion of one seed can probably kill a child.

Although all parts of the castor bean are actually poisonous, it is the seeds that are the most toxic. Castor bean ingestion causes a rapid onset of nausea, vomiting, stomach cramps, and severe diarrhea, followed by vascular collapse. Death usually occurs on the third day in the absence of appropriate medical intervention.

Ricin is least toxic by the oral route. Signs and symptoms appear 4 to 8 hours after exposure.

Signs and symptoms of ricin ingestion are as follows:
- Fever
- Chills
- Headache
- Muscle aches
- Nausea
- Vomiting
- Diarrhea
- Severe abdominal cramping
- Dehydration
- Bleeding in the stomach and colon
- Death of liver, spleen, kidneys, stomach, and colon

Inhalation of ricin causes nonspecific weakness, cough, fever, hypothermia, and hypotension. Symptoms occur about 4 to 8 hours after inhalation, depending on the inhaled dose. The onset of profuse sweating some hours later signifies the end of the symptoms. Call EMS immediately.

Signs and symptoms of ricin inhalation are as follows:
- Fever
- Chills
- Nausea
- Local irritation of eyes, nose, and throat
- Profuse sweating
- Headache
- Muscle aches

- Nonproductive cough
- Chest pain
- Difficulty breathing
- Fluid in the lungs
- Severe lung inflammation
- Convulsions
- Breathing failure

6

Nuclear/Radiological Terrorism

TACTICAL ACTIONS

- Observe explosive protection procedures from radiological dispersion devices (RDD).
- Use the principles of time, distance, and shielding for radiation protection.
- Call for immediate assistance from local emergency response agencies.
- Immediately secure a radiological site.
- Decontamination is required after any suspected exposure.
- Assist local law enforcement in scene control and perimeter security if directed.

Introduction

There have been only two incidents involving the use of a nuclear device. During World War II, Hiroshima and Nagasaki were devastated when they were targeted with nuclear bombs. The awesome destructive power demonstrated by the attack ended World War II and has served as a deterrent to nuclear war.

Today, there are nations that hold close ties with terrorist groups (known as state-sponsored terrorism) and have obtained some degree of nuclear capability. It is also possible for a terrorist to secure radioactive materials or waste to perpetrate an act of terror. Such materials are far easier for the determined terrorist to acquire and would require less expertise to use. The difficulties in developing a nuclear weapon are well documented. Radioactive materials, however, such as those in Radiological Dispersal Devices (RDDs), also known as "dirty bombs," can cause widespread panic and civil disturbances.

What Is Radiation?

Ionizing radiation is energy that is emitted in the form of rays or particles. This energy can be found in radioactive material, such as rocks and metals. Radioactive material is any material that emits radiation.

The energy that is emitted from a strong radiological source is either alpha, beta, gamma (X-rays), or neutron radiation. Alpha is the least harmful type of radiation and cannot travel fast or through most objects. In fact, a sheet of paper or the body's skin easily stops it. However, alpha particles can cause serious injury or death if ingested. Beta radiation is slightly more penetrating than alpha, and requires a layer of clothing stop it. Gamma or X-rays are far faster and stronger than alpha and beta rays. These rays easily penetrate through the human body and require either several inches of lead or concrete to prevent penetration. Neutron energy is the fastest moving and most powerful form of radiation. Neutrons easily penetrate through lead and require several feet of concrete to stop them.

Sources of Radiological Material

There are thousands of radioactive materials found on the earth. These materials are generally used for purposes that benefit humankind, such as medicine, killing germs in food (irradiating), and energy. Once radiological material has been used for its purpose, the material remaining is called radiological waste. Radiological waste remains radioactive, but has no more usefulness. These materials can be found at:

- Hospitals
- Colleges and universities
- Chemical and industrial sites

Not all radioactive material is secured, and the waste is often not guarded. This makes use of radioactive material and substances appealing to terrorists.

Radiological Dispersal Devices (RDD)

A radiological dispersal device (RDD) is any container that is designed to disperse radioactive material. This generally requires the use of a bomb, hence the nickname "dirty bomb." A dirty bomb carries the potential to injure victims with not only the radioactive material but the explosive material used to deliver it. Just the thought of an RDD creates fear in a population, and so the ultimate goal of the terrorist—fear—is accomplished. In reality, the destructive capability of a dirty bomb is limited to the explosives that are attached to it. Therefore, if the explosive is sufficient to kill 10 persons without radioactive material, it could kill 10 persons with the radioactive material added. There may be long-term injury and illness associated with the use of an RDD, not much more than the bomb by itself would create. In short, the dirty bomb is an ineffective WMD.

Nuclear Energy

Nuclear energy is artificially made by altering (splitting) radioactive atoms. The result is an immense amount of energy that usually takes the form of heat. Nuclear material is used in medicine, weapons, naval vessels, and power plants. Nuclear material gives off all forms of radiation including neutrons (the most deadly type). Like radioactive material, when nuclear material is no longer useful it becomes waste that is still radioactive.

Nuclear Weapons

The destructive energy of a nuclear explosion is unlike any other weapon in the world. That is why nuclear weapons are kept only in secure facilities throughout the world. There are nations that have ties to terrorists that have actively attempted to build nuclear weapons. Yet the ability of these nations to deliver a nuclear weapon, such as a missile or bomb, is as of yet, unproven. Complete mutual annihilation is also the deterrent, therefore, the likelihood of a nuclear attack is extremely remote.

Unfortunately, however, due to the collapse of the former Soviet Union, the whereabouts of many small nuclear devices is unknown. These small suitcase-sized nuclear weapons are called Special Atomic Demolition Munitions (SADM). The SADM, or "suit-case nuke," was designed to destroy individual targets, such as important buildings, bridges, tunnels, or large ships.

Symptoms

The effects of radiation exposure vary depending on the amount of radiation that a person receives and the route of entry. Radiation can be introduced into the body by all routes of entry as well as through the skin (irradiation). The victim can inhale radioactive dust from nuclear fallout or from a dirty bomb, or have radioactive liquid absorbed into the body through the skin. Once in the body, the radiation source will irradiate the person from within rather than from an external source (such as X-ray equipment). Some common signs of acute radiation sickness are listed in **TABLE 6-1.** Additional injuries that occur with a nuclear blast include thermal and blast trauma, trauma from flying objects, and eye injury.

Table 6-1 Common Signs of Acute Radiation Sickness

Low exposure	Nausea, vomiting, diarrhea
Moderate exposure	First-degree burns, hair loss, death of the immune system, cancer
Severe exposure	Second- and third-degree burns, cancer, death

Medical Management

Being exposed to a source of radiation does not make a victim contaminated or radioactive. However, when victims have a radioactive source on their body (such as debris from a dirty bomb), they are contaminated and must be initially decontaminated by a hazardous materials responder.

Protective Measures

There are no suits or protective gear designed to completely shield from radiation. Those who work in high-risk areas do wear some protection (lead-lined suits); however, this equipment is not readily available. The best way to protect yourself from the effects of radiation is to use time, distance, and to shield yourself.

Time. Radiation has a cumulative effect on the body. The less time that you are exposed to the source, the less the effects will be. If you realize that the victim is near a radiation source, leave the area immediately.

Distance. Radiation is limited as to how far it can travel. Depending on the type of radiation, often moving only a few feet is enough to remove you from immediate danger. Alpha radiation cannot travel more than a few inches.

Shielding. As discussed earlier, the path of all radiation can be stopped by a specific object. It will be impossible for you to recognize the type of radiation being emitted, or even from which direction it is coming. Therefore, you should always assume that you are dealing with the strongest form of radiation and use concrete shielding (such as buildings or walls) between yourself and the incident. The importance of shielding cannot be overemphasized. In one atomic test, a car was parked on the side of a house, opposite the direction of the oncoming blast. The house was completely destroyed, yet the car that was directly next to it sustained almost no damage.

Explosives

- Call for immediate help—describe the device.
- Evacuate the immediate area in accordance with guidelines (See Appendix D).
- Never touch, remove, or examine the device or package.
- Do not touch or examine a suspicious vehicle.
- Look (from a distance) for a secondary device.
- Assist in an area search if requested by on-scene law enforcement.

Introduction

During the 1990s, there were a number of catastrophic bombing incidents in the United States. The 1993 bombing of the World Trade Center in New York killed six, injured several thousand, and caused more than $750 million in damage. The bombing of the Murrah Federal Building in Oklahoma City in 1995 took the lives of 168 innocent people, injured

518 people, and caused $100 million in damage. Some bombers are able to elude police for years. Theodore Kacazynski, known as the "Unabomber" killed three people and injured 22 others with 16 package bombs over a period of 18 years. Eric Rudolph, convicted for the Olympic Park bombing during the 1996 Olympic Games in Atlanta, eluded an intensive law enforcement manhunt until 2003. Improvised explosive devices (IED) are common terrorism weapons throughout the world.

The possibility that you would encounter an explosive device continues to increase. If you assume that a harmless-looking package, a normally safe area, or a calm, friendly individual will not present a threat, you invite disaster.

Explosives and Incendiary Devices

According to the FBI's Bomb Data Center, there has been a substantial increase in the number of bombings in the United States over recent years. Between 1987 and 1997, the FBI recorded 23,613 bombing incidents, resulting in 448 deaths and 4,170 injuries. Incendiary devices account for 20% to 25% of all bombing incidents in the United States.

Groups or individuals have used explosives to further a cause; to intimidate a co-worker or former spouse; to take revenge; or simply to experiment with a recipe found in a book or on the Internet.

Each year thousands of pounds of explosives are stolen from construction sites, mines, military facilities, and other locations. How much of this material makes its way to criminals and terrorists is not known. Terrorists can also use commonly available materials, such as a mixture of ammonium nitrate fertilizer and fuel oil (ANFO), to create their own blasting agents.

An improvised explosive device is any explosive device that is fabricated in an improvised manner. Vehicle-borne improvised explosive devices (VBIED) are devices that are concealed in a vehicle. An IED or VBIED could be contained in almost any type of package from a letter bomb to a truckload of explosives. The Unabomber constructed at least 16 bombs that were delivered in small packages through the postal service. The bombings of the World Trade Center in 1993 and the Alfred E. Murrah Federal Building in 1995 both involved rental trucks loaded with ANFO and detonated by a simple fuse.

Pipe Bombs

The most common improvised explosive device is the pipe bomb. A pipe bomb is simply a length of pipe filled with an explosive substance and rigged with some type of detonator. Most pipe bombs are simple devices, made with black powder or smokeless powder and ignited by a fuse. More sophisticated pipe bombs may use a variety of chemicals and incorporate electronic timers, mercury switches, vibration switches, photocells, or remote control detonators as triggers.

Pipe bombs are sometimes packed with nails or other objects so they will inflict as much injury as possible on anyone in the vicinity. A chemical or biologic agent or radiological material could be added to a pipe bomb to create a much more complicated and dangerous incident. Experts can only speculate at the number of casualties that could result from such a weapon.

Suicide Bombings

Suicide bombings are an effective tactic used throughout the world and are expected to migrate to the United States. These devices can be concealed in a vehicle or on an individual. There is no common age or ethnic profile for suicide bombers. In recent years, women have joined their ranks. Precautionary measures include:

- Look for signs of suspicious behavior.
- Note unusual dress such as winter coats in the summer.
- Attempt to stop suspicious vehicles entering a critical area.
- Avoid directly approaching suspicious people or vehicles—call for help.

Secondary Devices

Terrorists may have placed a secondary device in the area where an initial event has occurred. These devices are intended to explode some time after the initial device explodes. Secondary devices are designed to kill or injure emergency responders, law enforcement personnel, spectators, or news reporters. Terrorists may use this tactic to attack the best-trained and most experienced investigators and emergency responders, or simply to increase the levels of fear and chaos following an attack.

The use of secondary devices is a common tactic in incidents abroad and has occurred at a few incidents in North America. In 1998, a bomb exploded outside a Georgia abortion clinic. About an hour later, a second explosion injured seven people, including two emergency responders. A similar secondary device was discovered approximately one month later at the scene of a bombing in a nearby community. Responders were able to disable this device before it detonated.

Working with Other Agencies

Joint training with local, state, and federal agencies charged with handling incidents involving explosive devices should occur on a routine basis. Among these agencies are local and state police; the FBI; the Bureau of Alcohol, Tobacco, and Firearms; and military explosive ordnance disposal (EOD) units. This is a great opportunity for the security facilities officials to gain a better understanding of emergency responders and law enforcement's needs. These forums are also ideal for sharing any unique needs of the security facilities official.

Potentially Explosive Devices

When you are faced with a bomb scare or suspicious package with a possible explosive, locate a knowledgeable representative from the area. This may be a family member if the incident is at a residence or a business representative if it occurs at a commercial or industrial establishment.

A form developed by the FBI contains pertinent questions to ask and information to gather during a bomb threat (**FIGURE 7-1**). This information is critical in assessing the credibility of the threat. Unfortunately, most serious bombers rarely phone-in a threat to their targets.

Any information obtained by the person who takes the call will be useful. The expected time of detonation and the location of the device are the most important questions you should ask. If time permits, obtain floor plans of the building and make enough copies on a copy machine in a nearby building so that each search team has plans of its assigned search areas. Fire department preplans may also provide useful.

Before you issue an evacuation order, have the occupants examine the immediate area around them for anything suspicious. These individuals will have the greatest knowledge of their surroundings and will be able to spot an abnormal object. Instruct them not to touch anything suspicious, but to report their findings to on-scene emergency responders. To keep the panic at a minimum, officials may elect to quietly evacuate the building, perhaps on the pretense of a fire alarm. In this case, one key individual from the building would be used to search the premises. Evacuees should be kept at a safe distance from the scene. On-scene law enforcement personnel can assist in determining a safe distance. Remember that the range of flying glass and debris from a bomb will be far greater than you might expect.

FD-730 (Rev. 6-20-97)

FBI
BOMB DATA CENTER

Place This Card Under Your Telephone

QUESTIONS TO ASK:

1. When is the bomb going to explode?

2. Where is it right now?

3. What does it look like?

4. What kind of bomb is it?

5. What will cause it to explode?

6. Did you place the bomb?

7. Why?

8. What is your address?

9. What is your name?

EXACT WORDING OF THE THREAT:

_____

_____

_____

_____

_____

_____

_____

_____

Sex of caller: _____ Race: _____

Age: _____ Length of call: _____

Number at which call is received: _____

Time: _____ Date: _____

BOMB THREAT

FBI/DOJ

FIGURE 7-1A Form developed by the FBI contains pertinent questions to ask and information to gather during a bomb threat. (See Appendix D for more information.)

CALLER'S VOICE:

_____	Calm	_____	Nasal
	Angry		Stutter
_____	Excited	_____	Lisp
_____	Slow	_____	Raspy
	Rapid	_____	Deep
_____	Soft		Ragged
_____	Loud	_____	Clearing throat
	Laughter	_____	Deep breathing
_____	Crying	_____	Cracking voice
_____	Normal	_____	Disguised
_____	Distinct	_____	Accent
_____	Slurred	_____	Familiar
		_____	Whispered

If voice is familiar, who did it sound like?

_____

_____

BACKGROUND SOUNDS:

_____	Street noises	_____	Factory machinery
_____	Crockery	_____	Animal noises
	Voices	_____	Clear
_____	PA System	_____	Static
_____	Music	_____	Local
	House noises	_____	Long distance
_____	Motor	_____	Booth
_____	Office machinery	Other _____

_____

THREAT LANGUAGE:

	Well spoken (educated)		Incoherent
		_____	Taped
_____	Foul	_____	Message read by threat maker
_____	Irrational		

REMARKS: _____

_____

_____

_____

Report call immediately to:

_____

Phone number _____

· ·

Date _____ / _____ / _____

Name _____

Position _____

Phone number _____

FIGURE 7-1B Form developed by the FBI contains pertinent questions to ask and information to gather during a bomb threat.

8

Decontamination Techniques

TACTICAL ACTIONS

- Make every attempt to avoid being contaminated on an incident scene.
- Initiate immediate scene control procedures to ensure that contaminated personnel or victims are directed to a decontamination area.
- Notify emergency response teams.
- Do not enter contaminated areas ("hot zones") without personal protective equipment.
- If contaminated, remove clothing and shoes as soon as possible and exit the area through the decontamination corridor.
- All exposed security officers must be medically evaluated after decontamination.

What is Decontamination?

Decontamination is the physical or chemical process of reducing and preventing the spread of hazards by persons and equipment. Decontamination makes personnel, equipment, and supplies safe by removing or eliminating hazards. Proper decontamination is essential at every incident to ensure the safety of personnel and property.

Secondary Contamination

Contamination is the process of transferring a hazard from its source to people, animals, the environment, or equipment, which may act as carriers of the contaminant. Secondary contamination (also known as cross-contamination) occurs when a contaminated person or object comes into direct contact with another person or object. Secondary contamination may occur in several ways:

- A contaminated victim comes into physical contact with another person.
- A bystander comes into contact with a contaminated object.

Because secondary contamination can occur in so many ways, control zones should be established, clearly marked, and enforced at incidents. The more people who are exposed to the contaminant, either directly or indirectly, the larger the decontamination problem.

Safe and Initial Response Procedures

You should:

- Recognize the event as terrorism and a potential crime scene, preserving evidence where possible.
- Perform a safe approach to the site. Approach cautiously from uphill and upwind, if possible.
- Notify local law enforcement and emergency response agencies.
- Isolate the area, deny entry, and ensure scene security.
- Identify and address structural exposures. Shut down ventilation or product distribution points if you can do so safely.
- Monitor weather reports to ensure weather awareness, especially wind direction. Identify other environmental exposures and address these issues.

- Do not enter the hazard scene, even if victims are present. The victims could contaminate you, making you a victim yourself.
- Use disposable coveralls (for example, Tyvek™) to replace contaminated uniforms.
- Document all exposures and medical evaluations.

Emergency Decontamination

NOTE: This section is for security officer overview and familiarization. Decontamination operations should only be conducted by properly trained personnel with full personal protective equipment.

Emergency decontamination is used in potentially life-threatening situations to rapidly remove most of the contaminants from an individual, regardless of a formal decontamination corridor (a controlled area where decontamination procedures take place). A more formal and detailed decontamination process may follow later.

Emergency decontamination usually involves removing contaminated clothing and dousing the victim with quantities of water. If a decontamination corridor has not yet been established, isolate the exposed victims in a contained upwind and uphill area and coordinate with emergency responders to establish an emergency decontamination area. If possible, do not allow the water runoff to flow into drains, streams, or ponds. Try to divert it into an area where it can be treated and/or disposed of later. Do not delay decontamination; human life always comes first.

Crime Scene Awareness

TACTICAL ACTIONS

- Call for local law enforcement and initiate crime scene perimeter control immediately.
- Maintain situational awareness on the perimeter and interior of a crime scene.
- Coordinate evidence management with local law enforcement.
- Remember that all individuals on the scene are potential witnesses or perpetrators.

Introduction

The increase in violent crime also increases the contact that security officers have with the victims of crime. Security officers have a personal responsibility to ensure their personal safety, as well as a responsibility to the community at large. By assisting law enforcement to maintain the integrity of the crime scene, security officers increase the probability that a suspect will be captured and convicted. That, in turn, reduces the possibility of future danger from additional crime.

Recognizing Your Incident as Terrorism

Security officers should be aware of the indicators of a potential terrorist event and initiate control actions if these indicators are observed. Be alert for actions directed against responders such as secondary devices, people, or vehicles. Implement safe initial response procedures to a terrorist incident.

Indicators that you may be encountering terrorist activities include:

- Vehicle ownership or rental car information that is contradictory, vague, or does not match other documentation.
- An individual who does not know where he or she is going and has no reason for being in the area.
- An individual carrying suspicious packages.
- Expired passport, visa, or another personal identification paper that is vague, altered, or false.
- Items related to infrastructure such as maps, photographs, or diagrams of potential targets, especially if they show construction/operational details.
- Persons taking pictures of a target facility or of security or ventilation equipment.
- Suspicious items in a vehicle or on a person, such as bomb-making materials, weapons of an unusual nature, or manuals for the manufacture of these items.
- Persons asking questions on security-related issues.
- Persons who exhibit unusual behavior that attracts attention or that can be considered threatening.
- Transportation of hazardous materials in an unconventional manner or without proper paperwork.

The following are some general signs of a terrorism incident where chemical, biological, radiological, nuclear, or explosive agents have been used:

- Disruptions to critical systems (such as transportation or utilities).
- Receipt of a threat.

- An explosion, particularly if there is a debris field.
- A secondary attack/explosion.
- Multiple non-trauma related victims.
- Responders who are victims.
- Hazardous substances involved in the incident.
- Severe structural damage without an obvious cause.
- Dead animals or vegetation.
- Unusual odors, color of smoke, or vapor clouds.

Signs found in victims indicating a terrorism incident where weapons have been used include:

- Unconscious victims with minimal or no trauma.
- Victims exhibiting SLUDGEM signs/symptoms:
  - Salivation, Sweating
  - Lacrimation (excessive tearing)
  - Urination
  - Defecation, Drooling, Diarrhea
  - Gastric upset and cramps
  - Emesis (vomiting)
  - Muscle twitching
- Victims having difficulty breathing.
- Blistering, reddening, discoloration, or irritation of the skin.
- Signs/symptoms common among multiple victims.

Safe Initial Response Procedures

Your primary response actions to a potential terrorism incident are life safety, incident stabilization, including crime scene security and environmental and property protection. You should:

- Recognize the event as terrorism and a potential crime scene, preserving evidence where possible.
- Perform a safe approach to the site. Approach cautiously, from uphill and upwind.
- Assess security.
- Implement public protection measures, including downwind evacuations, based on the severity of the incident.

- Isolate area, deny entry, and ensure scene security. Do not move or disturb items at the scene unless this is necessary for life-safety purposes. Photograph and document scene conditions, if possible.
- Identify safe and secure staging location(s) for incoming emergency response. Keep in mind that responders may be targets of terrorists.
- Identify and address structural exposures. Shut down ventilation or product distribution points, if you can do so safely.
- Be aware of weather conditions, as they may cause downwind or other exposures.
- Field-interview victims and witnesses if requested by local law enforcement and document the following:
  - Is everyone accounted for?
  - What happened?
  - When did it happen?
  - Where did it happen?
  - Who was involved?
  - Did they smell, see, taste, hear, or feel anything out of the ordinary?
  - Did they see suspicious packages, devices, vehicles, or people?

What Is Evidence

The collection and preservation of evidence is paramount for any successful criminal investigation. Without the appropriate and legal assembly of all the evidence and information surrounding a crime, the entire effort that follows could be for naught. The collection and preservation of evidence is the record of what occurred at a particular time and location— and what actions were taken by specific individuals. Altering the physical make-up of items could, possibly forever, alter the record of what occurred. As a result, investigators may lose their chance to identify suspects, conspirators, and witnesses and to secure convictions in a court of law. Sloppiness, inattention to detail, laziness, and lack of training can result in the denial of justice. *It is essential that all evidence management procedures be approved and coordinated with local law enforcement.*

Physical evidence generally falls into three categories: body materials, objects, and impressions. Body materials include fluids and materials such as blood (either fresh or dried), semen, tissue, hair, earwax, and skin. Objects include weapons, soil, tools, and documents. Impressions are not feelings, but forms or figures left by physical contact. For example, a single match lying beside a victim could be compared to the torn edges of a matchbook found in a suspect's pocket, linking him to the scene. The impression of a shoe print may link a suspect to a homicide or determine whether a driver had his foot on the brake pedal or the gas pedal at the time of an accident.

Crucial Precautions Upon Arrival at a Crime Scene

As a security officer responding to a violent crime scene, you must be concerned with your personal safety first. If you arrive on the scene of an explosion or terrorist event, coordinate with law enforcement to secure the area before you enter.

As you approach the crime scene, be mindful of bullet casings, weapons, blood spatter, or puddles. Law enforcement agencies may have a clear pathway void of evidence that you can use. All responders should use the same route to access and to leave the scene.

Open doors at a possible crime scene by using your shoulder, elbow, or the back of your hand. Even if you are wearing gloves, grabbing a door knob with your hand could destroy any fingerprints on the knob. Once inside the scene, try to disturb as little as possible. Review the scene. Is the door open or closed? Are the lights on or off? If you arrive at an unknown crime scene before law enforcement, write precise notes to document critical points.

Security officers may not be experts in evidence collection, but their prompt scene preservation will enhance the efforts of specialized follow-up personnel. Try to:

- Ensure that there are no imminent threats or hazards to health and safety of people in the immediate area.
- Identify and secure the boundaries of the crime scene by roping off or otherwise preventing unnecessary activity and traffic, including any area that is likely to yield evidence.
- Identify and protect secondary scenes, which are all avenues that could be utilized to travel to and from the primary scene.
- Do not allow unauthorized persons to enter the crime scene area.
- Establish only one point of entrance and exit.
- Prevent destruction or contamination of evidence.

- Photograph the crime scene to record its appearance.
- Record the weather, lighting, and environmental conditions at the crime scene. Note the conditions and time of changes if appropriate.
- Do not use the telephone.
- Do not shut down or turn off computers.
- Do not use bathroom facilities.
- Do not allow food or drinks to be brought into the crime scene.
- Encourage witnesses and bystanders to remain at the scene in safe areas until investigators have interviewed them.
- Do not allow the media to enter the crime scene—interviews and statements should be coordinated by the law enforcement Public Information Officer or facility managers.

Quick Reference Chemical Agents

Indicators of a Possible Chemical Incident

Dead animals/birds/fish	Not just an occasional road kill, but numerous animals (wild and domestic, small and large), birds, and fish in the same area.
Lack of insect life	If normal insect activity (ground, air, and/or water) is missing, check the ground/water, surface/shore line for dead insects. If near water, check for dead fish/aquatic birds.
Unexplained odors	Smells may range from fruity to flowery to sharp/pungent to garlic/horseradish-like to bitter almonds/peach kernels to new mown hay. It is important to note that the particular odor is completely out of character with its surroundings.
Unusual numbers of dying or sick people (mass casualties)	Health problems including nausea, disorientation, difficulty in breathing, convulsions, localized sweating, conjunctivitis (reddening of eyes/nerve agent symptoms), erythema (reddening of skin/vesicant symptoms), and death.
Pattern of casualties	Casualties will likely be distributed downwind, or if indoors, by the air ventilation system.
Blisters/rashes	Numerous individuals experiencing unexplained water-like blisters, weals (like bee stings), and/or rashes.

Indicators of a Possible Chemical Incident, continued

Illness in confined area	Different casualty rates for people working indoors versus outdoors dependent on where the agent was released.
Unusual liquid droplets	Numerous surfaces exhibit oily droplets/film; numerous water surfaces have an oily film (No recent rain.)
Different looking areas	Not just a patch of dead weeds, but trees, shrubs, bushes, food crops, and/or lawns that are dead, discolored, or withered. (No current drought.)
Low-lying clouds	Low-lying cloud/fog-like condition that is not consistent with its surroundings.
Unusual metal debris	Unexplained bomb/munitions-like material, especially if it contains a liquid.

Source: Emergency Response Guidebook 2004, US Department of Transportation, Transport Canada, and the Secretariat of Communications and Transportation of Mexico, 2004.

Chemical Agents

Name	Military Designation	Odor	Lethality	Onset of Symptoms	Volatility	Primary Route of Exposure
Nerve agents	Tabun (GA); Sarin (GB); Soman (GD); VX	Fruity or none	Most lethal chemical agents; can kill within minutes; effects are reversible with antidotes	Immediate	Moderate (GA, GD); Very high (GB); Low (VX)	GA-both; GB-vapor hazard; GD-both; VX-contact hazard
Vesicants	Mustard (H); Lewisite (L); Phosgene oxime (CX)	Garlic (H); Geranium (L)	Cause large blisters to form on victims; may severely damage upper airway if vapors are inhaled; severe, intense pain and grayish skin discoloration (L and CX)	Delayed (H); Immediate (L, CX)	Very low (H, L); Moderate (CX)	Primarily contact with some vapor hazard
Pulmonary agents	Chlorine (CL); Phosgene (CG)	Bleach (CL); Cut grass (CG)	Causes irritation choking (CL); severe pulmonary edema (CG)	Immediate (CL); Delayed (CG)	Very high	Vapor hazard
Cyanide agents	Hydrogen cyanide (AC); Cyanogen chloride (CK)	Almonds (AC); Irritating (CK)	Highly lethal chemical gases; can kill within minutes; effects are reversible with antidotes	Immediate	Very high	Vapor hazard

Isolation and Evacuation Distances: Chemical Agents

Agent	Initial Isolation (All Directions) Small Release or Package (Feet)	Initial Isolation (All Directions) Large Release or Multiple Small Package (Feet)	Downwind Daytime Evacuation Small Release or Package (Miles)	Downwind Daytime Evacuation Large Release or Multiple Small Package (Miles)	Downwind Night Evacuation Small Release or Package (Miles)	Downwind Night Evacuation Large Release or Multiple Small Package (Miles)
Metabolic or Blood Agents (Not otherwise identified.)	200	1500	0.4	2.5	1.5	5.0
Nerve Agents (Not otherwise identified.)	500	3000	1.0	7.0+	2.1	7.0+
Vesicants or Blister Agents (Not otherwise identified.)	300	2500	0.5	9.2	1.1	6.5
Cyanogen Chloride	200	1300	0.4	2.5	1.5	5.0
Hydrogen Cyanide	200	1500	0.1	1.0	0.3	2.4
Lewisite	100	300	0.1	0.6	0.2	1.1
Phosgene Oxide	100	300	0.1	0.6	0.3	1.9
Sarin	500	3000	1.0	7.0+	2.1	7.0+
Soman	300	2500	0.5	9.2	1.1	6.5
Sulfur mustard	100	200	0.1	0.4	0.1	0.7
Tabun	100	500	0.2	1.0	0.4	1.9
V agent	100	200	0.1	0.4	0.1	0.6

Source: *Emergency Response Guidebook 2004*, US Department of Transportation, Transport Canada, and the Secretariat of Communications and Transportation of Mexico, 2004.

B

Quick Reference Biologic Threats

Indicators of a Possible Biologic Incident

Unusual numbers of sick or dying people or animals	Any number of symptoms may occur. Casualties may occur hours to days after an incident has occurred. The time required before symptoms are observed is dependent on the agent used.
Unscheduled and unusual spray being disseminated	Especially if outdoors during periods of darkness.
Abandoned spray devices	Devices may not have distinct odors.

Source: Emergency Response Guidebook 2004, US Department of Transportation, Transport Canada, and the Secretariat of Communications and Transportation of Mexico, 2004.

Characteristics of Anthrax

Dissemination	Aerosol
Communicability	Only when passed through the skin (rare).
Route of entry	Through inhalation of spore of skin contact with spore or direct contact with skin wound.
Signs and symptoms	Flu-like symptoms, fever, breathing distress, and breathing failure after 3 to 5 days of flu-like symptoms.

Characteristics of Botulinum Toxin

Dissemination	Aerosol or food supply sabotage or injection.
Communicability	None
Route of entry	Ingestion or gastrointestinal
Signs and symptoms	Dry mouth, stomach blockage, urinary retention, constipation, nausea and vomiting, abnormal pupil dilation, blurred vision, double vision, drooping eyelids, difficulty swallowing, difficulty speaking, and breathing failure due to paralysis.

Characteristics of Plague

Dissemination	Aerosol
Communicability	Bubonic: low, only from contact with fluid in buboe (inflamed lymph glands). Pneumonic: high, from person to person.
Route of entry	Ingestion, inhalation, or skin contact.
Signs and symptoms	Fever, headache, muscle pain and tenderness, pneumonia, shortness of breath, extreme lymph node pain and enlargement (bubonic).

Characteristics of Smallpox

Dissemination	Aerosolized for warfare or terrorist uses.
Communicability	High from infected individuals or items (such as blankets used by infected victims). Person-to-person transmission is possible.
Route of entry	Through inhalation of coughed droplets or direct skin contact with blisters.
Signs and symptoms	Severe fever, body aches, headaches, small blisters on the skin, bleeding of the skin and nasal passages. Incubation period is 10 to 12 days and the duration of the illness is approximately 4 weeks.

Characteristics of Viral Hemorrhagic Fevers

Dissemination	Direct contact with an infected person's body fluids. It can also be aerosolized for use in an attack.
Communicability	Moderate from person to person, or contaminated items.
Route of entry	Direct contact with an infected person's body fluids.
Signs and symptoms	Sudden onset of fever, weakness, muscle pain, headache, and sore throat. All of these symptoms are followed by vomiting and as the virus runs it course, internal and external bleeding.

C

Quick Reference Radiological Incident

Indicators of a Possible Radiological Incident

Radiation symbols	Containers may display a "propeller" radiation symbol.
Unusual metal debris	Unexplained bomb/munitions-like material.
Heat-emitting material	Materials that is hot or seems to emit heat without any sign of an external heat source.
Glowing material	Strongly radioactive material may emit or cause radioluminescence.
Sick people/animals	In very improbable scenarios, there may be unusual numbers of sick or dying people or animals. Casualties may occur hours to days or weeks after an incident has occurred. The time required before symptoms are observed is dependent on the radioactive material used, and the dose received. Possible symptoms include skin reddening or vomiting.

Source: Emergency Response Guidebook 2004, US Department of Transportation, Transport Canada, and the Secretariat of Communications and Transportation of Mexico, 2004.

D

Quick Reference Explosives

Department of the Treasury
Bureau of Alcohol, Tobacco & Firearms
BOMB THREAT CHECKLIST

1. When is the bomb going to explode?

2. Where is the bomb right now?

3. What does the bomb look like?

4. What kind of bomb is it?

5. What will cause the bomb to explode?

6. Did you place the bomb?

7. Why?

8. What is address?

9. What is your name?

EXACT WORDING OF BOMB THREAT:

_____

_____

_____

_____

_____

Sex of caller: _____ Race: _____

Age: _____ Length of call: _____

Telephone number at which call is received: _____

Time call received: _____

Date call received: _____

CALLER'S VOICE

☐ Calm   ☐ Nasal

☐ Soft   ☐ Angry

☐ Stutter   ☐ Loud

☐ Excited   ☐ Lisp

☐ Laughter   ☐ Slow

☐ Rasp   ☐ Crying

☐ Rapid   ☐ Deep

☐ Normal   ☐ Distinct

ATF F 1613.1 (Formerly ATF F 1730.1, which still may be used) (6-97)

☐ Slurred   ☐ Whispered

☐ Ragged   ☐ Clearing Throat

☐ Deep Breathing   ☐ Cracking Voice

☐ Disguised   ☐ Accent

☐ Familiar *(If voice is familiar, who did it sound like?)* _____

BACKGROUND SOUNDS:

☒ Street noises   ☐ Factory machinery

☐ Voices   ☐ Crockery

☐ Animal noises   ☐ Clear

☐ PA System   ☐ Static

☐ Music   ☐ House noises

☐ Long distance   ☐ Local

☐ Motor   ☐ Office machinery

☐ Booth   ☐ Other *(Please specify)*

BOMB THREAT LANGUAGE:

☐ Well spoken (education)   ☐ Incoherent

☐ Foul   ☐ Message read by threat maker

☐ Taped   ☐ Irrational

REMARKS: _____

_____

Your name: _____

Your position: _____

Your telephone number: _____

Date checklist completed: _____

ATF F 1613.1 (Formerly ATF F 1730.1) (6-97)

Bomb threat checklist. Courtesy of the Department of the Treasury/ATF.

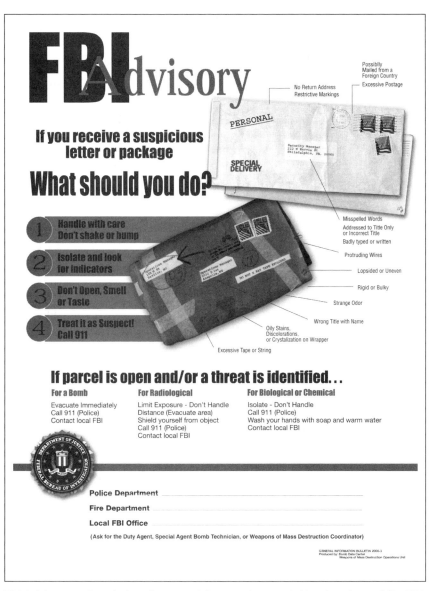

FBI Advisory on how to handle a suspicious package or letter. Courtesy of the FBI.

THREAT	THREAT DESCRIPTION	EXPLOSIVES CAPACITY[1] (TNT EQUIVALENT)	BUILDING EVACUATION DISTANCE[2]	OUTDOOR EVACUATION DISTANCE[3]
PIPE BOMB	5 LBS/ 2.3 KG	70 FT/ 21 M	850 FT/ 259 M
BRIEFCASE/ SUITCASE BOMB	50 LBS/ 23 KG	150 FT/ 46 M	1,850 FT/ 564 M
COMPACT SEDAN	500 LBS/ 227 KG	320 FT/ 98 M	1,500 FT/ 457 M
SEDAN	1,000 LBS/ 454 KG	400 FT/ 122 M	1,750 FT/ 534 M
PASSENGER/ CARGO VAN	4,000 LBS/ 1,814 KG	640 FT/ 195 M	2,750 FT/ 838 M
SMALL MOVING VAN/DELIVERY TRUCK	10,000 LBS/ 4,536 KG	860 FT/ 263 M	3,750 FT/ 1,143 M

This card supersedes any previous undated versions 11/99

Terrorist Bomb Threat Stand-Off

THREAT	THREAT DESCRIPTION	EXPLOSIVES CAPACITY[1] (TNT EQUIVALENT)	BUILDING EVACUATION DISTANCE[2]	OUTDOOR EVACUATION DISTANCE[3]
MOVING VAN/ WATER TRUCK	30,000 LBS/ 13,608 KG	1,240 FT/ 375M	6,500 FT/ 1,982 M
SEMI-TRAILER	60,000 LBS/ 27,216 KG	1,570 FT/ 475 M	7,000 FT/ 2,134 M

All personnel must evacuate (both inside of buildings and out).

All personnel must either seek shelter inside a building (with some risk) away from windows and exterior walls, or move beyond the Outdoor Evacuation Distance.

Preferred area (beyond this line) for evacuation of people in buildings and mandatory for people outdoors.

[1] Based on maximum volume or weight of explosive (TNT equivalent) that could reasonably fit in a suitcase or vehicle.
[2] Governed by the ability of an unstrengthened building to withstand severe damage or collapse.
[3] Governed by the greater of fragment throw distance or glass breakage/ falling glass hazard distance. Note that pipe and briefcase bombs assume cased charges which throw fragments farther than vehicle bombs.

Terrorist bomb threat stand-off card. Courtesy of TSWG (Technical Support Working Group)